Sewn Boats of Lakshadweep

Lotika Varadarajan

National Institute of Oceanography

Published by	National Institute of Oceanography
Printed at	UIC Printing & Packaging Pte Ltd, Singapore
Design	G V Soumitri, Sophie Gaur, Anurag Gupta
Print Production	Word Smithy
Price	Rs. 1500 US$ 100

ISBN 81-85121-29-X

Dedicated to the seafarers of Lakshadweep

Contents

Foreword

India is endowed with rich diversity of natural resources and of human resources with occupational skills that span the modern and traditional knowledge systems. The Council for Scientific and Industrial Research (CSIR) is committed to assist the nation in deriving enhanced and sustainable value from this entire spectrum of endogenous resource base. The traditional occupations practised by over four thousand communities in India are of particular significance in our search for sustainable development, as the underlying technological practices of these occupations have traditionally been tuned to ecological and territorial endowments of the area where these communities subsist. This rich cultural fabric of flexibly interconnected occupational skills, technology, artefacts and markets presents a fascinating and useful research and development agenda, an agenda that is inherently multidisciplinary.

In this context, CSIR, in collaboration with other agencies has nucleated studies and documentation of the rich techno-cultural heritage of India. This publication deals with the traditional technique of coir stitched boat building of the islands of Lakshadweep. The different stages in *Odam* building have been documented in detail. Observation was made of activities such as sea and pit retting of coir, manual rope making for Odam sewing, removal of copra from coconut, meera tapping, lime making, cowrie gathering and lagoon fishing, including techniques of spearing. Island culinary delicacies, rarely available to the passing traveller owing to the non-commercial nature of production, showed how deeply the immediate environment had pervaded local life in the past. The rhythm of daily life and details of house construction and usage of human muscle power as energy source were also observed and noted. All these have highlighted the cyclical link between man and environment free of the factor of pollution fallout. The life sustaining and environmentally friendly properties of marine life and the coconut palm were demonstrated in many different ways. The report reveals the gamut of skills brought into operation consequent to constraints of a coral island habitat.

I believe that with this book an important beginning has been made towards making known the rich tradition of boat building in India and for this I would particularly like to express my grateful thanks to Dr. Lotika Varadarajan.

New Delhi
July 1998

Dr. R.A. Mashelkar
Director General
Council for Scientific and Industrial Research

Acknowledgments

This book is based on a project assigned to NISTADS by the Department of Culture, Ministry of Human Resource Development, Government of India. The immediate objective of the project was to document the traditional technique of coir stitched boat building of the islands of Lakshadweep.

In Lakshadweep the project was coordinated by the Department of Social Welfare, Kavaratti. All help was provided by the Lakshadweep Administration and by Mr. Atta Koya Thangal, Secretary, Pradesh Council deputed by the Lakshadweep Administration to organise the schedule of the tour programme of Dr. Lotika Varadarajan deputed by NISTADS to conduct the study. The insights of Mr. Atta Koya Thangal have enriched this project beyond measure. Mr. Vijay Anand, Senior Research Fellow, Central Marine Fisheries Research Institute (CMFRI), Cochin, undertaking field investigations in Kavaratti, accepted the assignment of photographic documentation. Inter-island transport comprised speedboat, country boat and ship.

The following persons were contacted at the different islands:

Kavaratti: Kutti Yammad Bandeka Kada, master carpenter; Kunhi Kunhi Maistry, navigator, Hamsath Vellipura, Kala Illam, *terrapam* owners; K. Muhammad Haji Kunnel, master boat builders; Aboobacker, Department of Fisheries; Hanif, artist and sculptor.

Chetlat: Kunnel Shaikh, boat builder; Ayshachatta Ibrabim Kutty Haji, boat owner and navigator; Kunhi Ahmad, boat builder.

Kadmat: Charuthotham Zachariah, performing artist; U.C. Syed Ukkaya Chetta, boat builder.

Amini: Awwakkal Hamza, boat owner; Pura Ahmad Haji, also called Akkaka, boat builder.

Andrott: M.N. Koyamma Haji, boat builder; Mapillate Pukoya, navigator, Kunnakalam Cheriya Koya Thangal, community leader; the Principals of the Government High School and M.G. College .

Kalpeni: Taithotham Atta Koya, K.P. Pookoya, Kunnipoovakada, boat builders; Dr. K.K. Muhammad Koya, archaeological information;

Minicoy: Fatima Manika, Muraduganduwar, for details of Minicoy social structures; Ismail Bandarage, boat builder; Musa Manikfan, Bauge Meduge, navigator; Ali Ismail, Hindi teacher and Khatib, for historical details; Ali Thakru, Edurukage, traditional fisherman; Hasan Dombadage, sailor.

The grant given by the Department of Culture for carrying out this study is gratefully acknowledged.

Our sincere thanks to Dr. Lotika Varadarajan for conducting the study and for writing the book with great enthusiasm despite all the difficulties and extreme pressure on her time.

We record our appreciation of the painstaking task of formatting, designing and printing the book undertaken by Mr. G.V. Soumitri, Ms. Sophie Gaur and Mr. Anurag Gupta of Word Smithy, New Delhi.

Finally our thanks to all those in Lakshadweep, NISTADS and NIO for extending cooperation and help in bringing out the book.

A joint effort of the National Institute of Oceanography (NIO) and the National Institute of Science Technology and Development Studies (NISTADS), constituent establishments of the Council of Scientific and Industrial Research (CSIR), this book is expected to be followed by others in the series 'Indian Tradition of Navigation in the Indian Ocean'.

Ehrlich Dessa
Director
NIO

Ashok Jain
Director
NISTADS

July 1998

Introduction

Introduction

The Indian Ocean coral islands which constitute the Union territory of Lakshadweep may be seen to fall into two distinct entities. The units earlier called Amindivi (comprising Amini, Kadmat, Chetlat, Kiltan, Bitra), and Laccadives (including Kavaratti, Agatti, Kalpeni, Andrott), fall into one while the lone island of Minicoy constitutes the second. There are 10 inhabited and 26 uninhabited islands. On 1 November, 1956, these islands were brought together to form a Union Territory, the name Lakshadweep being conferred on it in 1973. For purposes of convenience, the terms Amindivi and Laccadives, will be retained in reference to the two island clusters mentioned above constituting the first unit. Prior to 1961 all the islands were classified as urban. At present Amini, Kavaratti and Minicoy have been categorised as urban, the other islands as rural. The 1981 Census has tabulated 93% of the population as Scheduled Tribe.

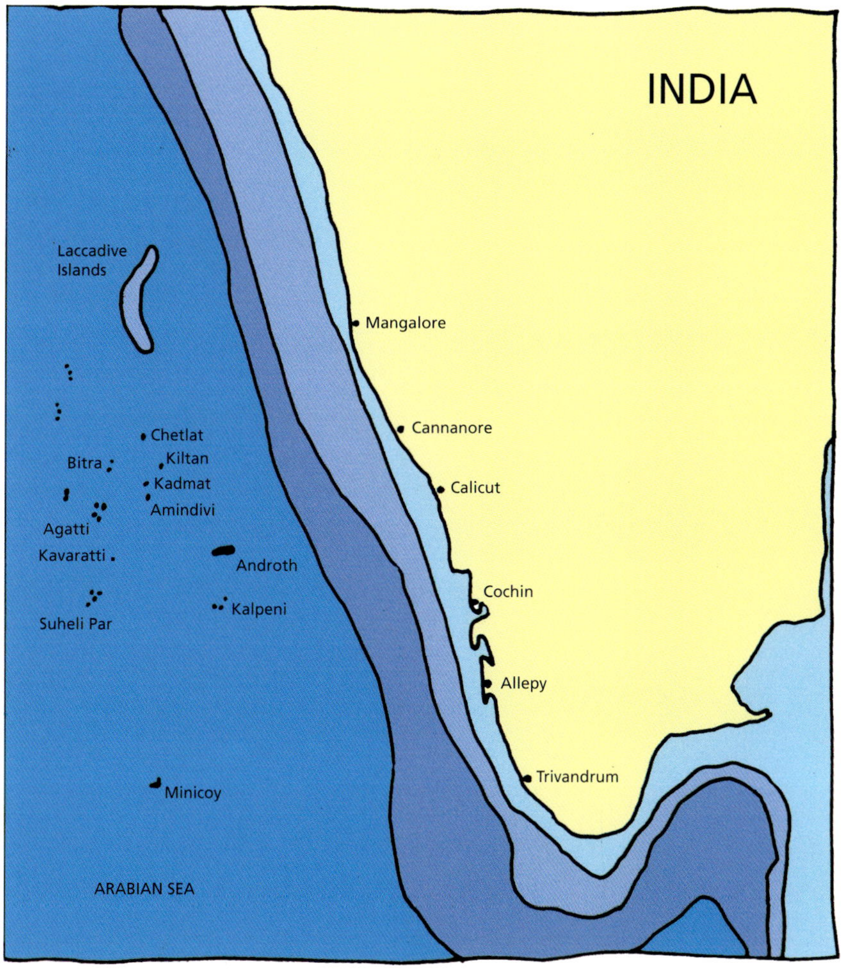

Historical Background

Since these islands do not have any aboriginal groups it is difficult to estimate when these may have been first populated, but, with the exception of Minicoy, the parent stock of the present inhabitants appears to have been very largely derived from Kerala. The impact of Sri Lanka may perhaps be studied in Minicoy, an extension of Maldivian culture, rather than the other islands. In view of the islamicisation of the islands Arab ethnicity cannot be ignored particularly on islands such as Andrott *(fig. 1)*. On the basis of linguistic evidence it would appear that the progenitors of the present group of settlers first stationed themselves on the so-called Tarwad islands of Amini, Kalpeni, Andrott, Kavaratti and Agatti at a time when Malayalam was still closely linked with Tamil on the mainland. The upper classes of the first four islands claim descent from Nayar, the warrior group, and Nambudris, Brahmins of the highest status in Kerala. The reference point of Nambudri, on the islands may be regarded as a quest for legitimisation as it is unlikely that this group would have had cause, except for reasons of social ostracism, to migrate. Kiltan, Chetlat and Kadmat appear to have been peopled from Amini by the least privileged group, the Melacheris. The term *Melacheri*, it has been suggested, has its origin in the fact that in some islands the lower castes lived in the western part of islands and hence came to be called *mela*-western, *cheri*-part. Bitra was populated by settlers from Chetlat in 1928. The population would, therefore, also be predominantly Melacheri.

1. At Boat Jetty, Andrott.
Note features of islander on the right.

In the 11th century A.D., Valabha 11 of the Mushaka dynasty in Kerala is said to have annexed the islands. The reason may have been anxiety to milk the trade in coir. Island coir was highly prized throughout the sphere of boat building activity along the Arabian Sea littoral. The Portuguese had taken note of the coir industry on the West Coast in the sixteenth century (Garcia da Orta, *Colóquios dos Simples e Drogas da Índia, I, Lisboa, 1891, reprint 1987, 236-253)*, but it is not clear when the coir industry, as distinct from coconut as source of copra, was developed in Kerala. According to K.M. Pannikar (*A History of Kerala,* Annamalainagar, 1959, 177-178), the Portuguese served as the catalyst for the development of coir as a trade item in Kerala. The Portuguese introduced seed nuts of a larger type of coconut from Africa, possibly from the Canary Islands, the local name

2. Coir being retted in Kerala backwaters. Cannanore.

being *Kappal Thengai,* coconut (brought by) ships. However, for purposes of sewing and servicing the needs of naval gear, the sea retted coir of the islands was far superior to the brackish backwater retted items of Kerala (*fig. 2*). European records of the Company period make no reference to sea retted coir produced in the vicinity of Trivandrum, and the rise of the industry appears to have been a post seventeenth century phenomenon.

In the 14th century the Mushaka territory of Kasargod-Chirakkal came to be known as Kolathunad under a new reigning dynasty, the Kolathiris. In the mid-sixteenth century, the Kolathiri transferred these islands in *Jagir* to the hereditary Admiral, the Ali Rajas of Cannanore (*Azhi Raja/King of the Ocean),* subject to the payment of an annual sum of 6,000 *fanams.* The Ali Rajas had embraced Islam in the 12th-13th century A.D. They were known after their family name Arakkal and had their palace in Cannanore. After the mid-eighteenth century Arakkal asserted its independence from Kolathiri. At the same time all foreign ships were banned from putting in at Lakshadweep in order to ensure the hold of Cannanore-Arakkal on the islands. Under the new dispensation if the ruler was a woman she was called Arakkal Bibi, males being referred to as Ali Raja. Circa A.D. 1760 coir and coconuts were constituted into royal monopolies the islands being provided with supplies of rice in return. In 1787 Amindivi passed under the control of Tipu Sultan of Mysore, and, after his fall, the English East India Company brought it within the jurisdiction of the South Kanara District. This was a purely administrative measure and did not affect the linguistic or religious character of this unit. Cannanore submitted to the English in 1790 but British control over, Amindivi and Laccadives was fully established only in 1908.

The Maldives comprise a separate culture zone. Unlike Lakshadweep, there is a memory of an aboriginal nucleus in the Maldives, those found in modern times being called Giraavaru. The earliest settlers, probably from Sri Lanka, appear to have reached the islands by the 5th century B.C. Minicoy formed part of the Maldives but by the early 16th century it had been ceded to Cannanore. According to T.P.C. Gabriel (*Lakshadweep,* New Delhi, 1989, 66-67), it is possible that Bodu Takrufan of the Maldives donated Minicoy to Cannanore in 1573. H.C.P. Bell, (*The Maldive Islands,* Colombo, 1940, 92), however, asserts that the Ali Rajas took it by force. As had happened with regard to Laccadives and Aminidivi, Minicoy was also brought within the compass of British India from the year 1908.

Islam in Lakshadweep

Islam came to the Maldives in A.D. 1153 while in Lakshadweep the process reached completion in the 13th century A.D. Although no in-depth archaeological explorations or excavations have taken place, the discovery of votive objects, sculpture and architectural fragments has been reported in course of construction activity. The best preserved are the Buddhist sculptures and assorted remains found in Andrott (*figs. 5-8*). Sri T. Satyamoorthy, Director, Art Museum, Trivandrum, has dated the Buddha head (*fig. 5*) to the 12th century A.D., assigning it a provenance of Sri Lanka, but the last word has yet to be pronounced on the subject. Figures of deities in the Hindu pantheon are yet to be identified.

3. Children doing homework. Government School, Chetlat.

4. Hujra Mosque, Kavaratti. Ornamental woodwork executed by local carpenter

5. Buddha head. Sandstone and plaster.

6. Torso. Coral. Front view.

7. Animal head. Coral.

8. Animal head. Coral. Figs 5-8, Courtesy, Government High School, Andrott.

There is also an oral tradition that the saint, Obaidullah, converted the islanders to Islam circa 41-42 A.H./661 - 662 A.D. (*fig. 9*). Professor S.Q. Fatimi, Islamabad, Pakistan, in a personal communication has tried to identify Obaidullah with Abu Talha, a contemporary of the Prophet, who died at sea circa 51 A.H./671 A.D. and was said to have been buried on an island. According to S.Q. Fatimi, the name *Abu Talha* transited through *Ubayd Allah* and *Ubaydullah* to *Obaidullah*. This would place conversion on the islands to the period of the first wave of proselytisation in Islam rather than to the second, as is currently assumed. In this context some attention has to be paid to the recent find of a tombstone within the precincts of Sheikh Palli at Agatti. If the script on the tombstone is read as Arabi-Malayalam there are no problems and the date provided is 139 A.H./759 A.D. If the script is taken as Arabic it yields the date 513 A.H./1133 A.D. (*fig. 10*). There are problems from either set of readings. The earliest Arabi-Malayalam text on the mainland is dated to the early 16th century. Moreover, if the early date is accepted the script should be Kufic. Scholars of Arabi-Malayalam on the islands have no difficulty in reading the inscription as Arabi-Malayalam, whereas if it is read as Arabic, interpolations have to be made. Other tombstones also exist exhibiting the same discrepancy between imputed date and visible script (*fig. 11*). One interpretation is that the first numerical indication of a four digit calendar year had been omitted as it was assumed that this would be taken for granted by contemporary readers. If this is accepted as a customary practice there would be no problem in finding a concordance between the abbreviated date and the script. However, the paleography is of a much later date and this discordance cannot be explained at this moment of time.

9. Small shrine to commemorate preaching of Obaidullah, Andrott. Note typology of oil lamp.

Leaving this problem aside, in view of the observations of S.Q. Fatimi, the oral tradition which dates the conversion of the islands to the seventh century A.D., deserves serious consideration. In the Maldives, because of closer links with Sri Lanka and the larger stretch of islands the course of islamicisation may have taken longer. As has been pointed out by Hasan Maniku (Male, Maldives), the new religion may have proved all the more acceptable as it provided a means for this island group to distance itself from Sri Lanka. The thesis advanced by A.D.W. Forbes (*Archipel*, 1981, 55-92), that Islam was introduced not from

Malabar, but from Yemen and the Hadramaut, may be accepted in relation to the Maldives but not Lakshadweep. The admixture of Arab blood in the population of the latter appears confined to a few leading families in Kalpeni and Andrott. On the other hand, social conventions, dress and the position accorded to Thangals within the community all point to the Mappilas of Malabar as progenitors of present day Islam in Lakshadweep.

10. Tombstone found at Sheih Palli, Agatti.

11. Tombstone located at *maqbara* of Obaidullah, Andrott. Imbibhj Koya, Chief Qazi, Calicut, reads the date as A.H. 300 but agrees that the digit indicative of the year, one thousand, has been omitted.

Language and Script

14.

The root language of all islands of Lakshadweep, with the exception of Minicoy, is Malayalam. Vaṭṭeluṭṭu was the earliest script used. It contained a heavy Sanskrit component. With the translation of Bhakti texts modern scripts based on Grantha were developed in the south. By the 16th century, higher castes among the Hindus had given up usage of Vaṭṭeluṭṭu and by the 17th century, modern Malayalam, *aryaneluṭṭu,* had attained its present from. However, as *Koleluṭṭu* sceptre writing, the earlier script was used for purposes of the framing of royal grants. Christians and Muslims, in the meanwhile, continued to use the Vaṭṭeluṭṭu script and in Lakshadweep this system of autography is in evidence in local pilot's (*malmis*) sailing manuals, in inscriptions on tombstones and those in some mosques/*pallis*. With the introduction of Islam, Arabi-Malayalam, associated with the literature of the Mappilas, developed on the mainland, had also came into usage on the islands. Arabi-Malayalam was based on the Arabic script. The language was a compendium of loan words derived from Arabic, Persian, Tamil, Kanara and Malayalam, *(fig 12).* The earliest text in Arabi-Malayalam is dated in Kerala to the 16th century as has already been pointed out. Arabic texts are also available and esoteric references as in *abjad* are in evidence. In *abjad* letters of the alphabet are assigned specific numerical values and certain numerical notations can also be taken as text *(fig. 13*). The language and script of the Maldives is Divehi, that of Minicoy is Mahl, closely aligned with Divehi. Mahl is said to be akin to Divehi as spoken in the seventeenth century. There appears to have been some interaction with Christianity long before the appearance of the Portugese as is evidenced by scattered tombstones which evade explanation as of now.*(fig. 14*)

13. Tombstone. Inscription in *abjad*. Mohiuddin Mosque, Kalpeni.

12. *Rehmani*. Arabi-Malayalam. Owned by Kidave Maidanpura, Amini.

Socio-Economic Organization

The socio-economic organization of the islands shows that the absence of monetisation in the past had led to the emergence of varied patterns of exchange in service and kind between different individuals in consonance with their social status. Mappila imprint is very strong in the social organization of the Laccadives and Amindivi while the mores of the community have been shaped by the limited resources available in an island environment. Apart from subsistence fishing, all sections did not necessarily, have to live off the sea. In Minicoy, however, the importance given to seafaring has been reflected in their pattern of social organisation.

Manikfans and Thakrufans formed the privileged section of the community in Minicoy and they were not compelled to form part of any communal support system having sufficient resources to remain independent of this. Since Thakru and Raveri menfolk largely took to service at sea, they had to be away from the home front for extended periods. Boys were, therefore, placed under the supervision of the village headman, the *Bodukaka/Moopan (fig. 15)*, and girls under the female community leader, the *Bodudatha.(fig. 16)* After the age of twelve girls would report for village work organized by the *Bodudatha*. Boys were divided into groups according to age. Between 5 - 12 they were a part of the *Lavilikudeen*. The age group 12 - 18 were tutored by young men who had performed their first voyage to the mainland and were, therefore, entitled to wear a special

15. Far left: H. Hussain, Faleserry, Minicoy. Earlier served as *Moopan*.
16. Amina, *Bodudatha*, Faleserry, Minicoy.

kind of pantaloon called *handawal*. Girls got married at 15, young men at about 25. The latter could only do so if they had undertaken at least one voyage to the mainland.

A prominent aspect of societal formation of the islands is that although the whole population is staunchly Islamic both in Lakshadweep and the Maldives, the social configuration bears the stamp of an earlier caste oriented system. The Laccadives has a three tier social system comprising Koyas, Malmis and Melacheries, with the Melacheris being called Thandels in Kavaratti. In Amini, the terms used in the past were Tarwadi/Karnavar, Tanakampranaver, Kudyati and Melacheri. In Minicoy the social divisions comprise the Manikfans. Thakrufans, Thakrus and Raveris. Ownership of boats was confined to Koyas, Karnavars and Manikfans who tried to perpetuate their hold on the island economy by excluding others from the enjoyment of this privilege. Malmis and Thakrufans constituted the traditional classes from which pilots were drawn.

The occupation practiced by the Koyas was that of the management of *Otams/ Odams* and the conduct of trade in island products with the mainland. Malmis served as captains and navigators. Since Thangals, traditional healers present on the Tarwad islands, had to have a knowledge of astrological time keeping, they, along with members of the Malmi grouping, were called upon to fix the *muhurtham*, auspicious time, for the performance of various rites of passage.

17, 18. Tables and charts used by Thangals for predicting auspicious moments. Andrott. Etched by Kunnamkalam Muthu Koya Thangal. Turn of present century.

19. Cyclical calendar in disc form. *Maqbara* of Obaidullah, Andrott.

Astronomical and calendrical aspects of such activities have been found inscribed on wood and stone (*figs. 17-23*). It is of interest that the greatest *malmi*/pilot, and *ustad*/teacher who lived and practiced at Andrott at the turn of the century, was Kunnamkalam Muthu Koya Thangal (*cf. fig. 24*). This shows that while as a religious preceptor or head of a sect, a Thangal would not take to sea, other family members were free to practice as pilots. Thangals were accorded Koya status. Melacheri occupations included those of climbing and tapping of coconut trees.

20. Shadow time reckoner. Coral. Mohiuddin mosque, Kalpeni.

21.Detail of Fig 20. If one is facing the sun without looking at it, it is possible to assess when it is directly overhead. The square stone in fig 20 had markings on the surface earlier on also a vertical coral peg.

24. Kunnamkalam Cheriya Koya Thangal, Kadariya sect. Andrott. Grandson of Kunnamkalam Muthu Koyal Thangal.

22. Inscribed plank from mosque, Agatti. Shows times of prayer. Computed from 16 books which are cited. In the outermost circle malayalam months are marked followed by minutes and differences in time each day according to different seasons.

23. Sun-dial. Hujra mosque, Kalpeni. The shadow of the string will align with groove at noon. The dial has a north-south orientation.

A.D.W. Forbes has pointed out that in relation to Malabar, the Melacheris inherited the traditional Tiyyar tasks of coconut gathering and toddy tapping, combining these with the Mukkuvan function of fishing. The question of the origin of the Tiyyars and Mukkuvans in Kerala merits further study. Although associated with an original home in Sri Lanka, that of the Maldives cannot be ruled out.

Beating and twisting of coir, (except in the category of coir ropes used in boat building), was practiced by women of all groups while fishing and cultivation of land was open to all.

Exploitation by the privileged groups was expressed through the *Nadappu* form of tenancy which was articulated in terms of free service on board the ships sailing to the mainland and manipulation of freight dues paid in kind, apart from exploitative relations at the home base. Every attempt was made to prevent dilution of the right to own ships which could sail to the mainland. Those who did not belong to the traditional ship owning classes were proscribed from obtaining ship building materials from the mainland on the boats returning to the islands. Moreover, when the inhabitants of Amini claimed control over Kadmat they refused to allow the settlers to build their own boats. Because of control over ownership of *otams* there were no commercial boat building yards in Lakshadweep. Struggles to own *otams* by members of other groups is dated to the post 1869 period. In the early 20th century, Chetlat had emerged, but very much under the shadow of Amini, as a centre of boat building from where boats could be procured for other islands. The model of boats built for sailing to the mainland, the *Valia Odam*, appears to have been a variant of the Konkani *pattamar*.

Natural resources

Natural resources

Because of ecological constraints, natural resources were limited in the two island groups. Cultivation of crops like rice and sugarcane was restricted to a few islands on scarce sites such as those found at Andrott and Amini. The prime necessity was that of cereal for which breadfruit of the species *Artocarpus incisa*, could act as a partial substitute. Smoked skipjack/*bonito/Katsuwonus pelamys*, called *kalu bilamas*, in Mahl, and yellow fin tuna, *mas*, were traded against essential commodities in Minicoy and the Maldives. In the other islands of Lakshadweep, fishing was not a commercial activity in the same way as at Minicoy. This was because the fish which was locally consumed enjoyed no external demand. Fish in local demand included sun dried strips of *kotar*/sharkmeat, *parava*/flying fish/ *Exocoetus sp.* and seerfish/*Acanthocybium solandri*.

25. Money Cowry. Family Cypraeidae, species, Monetaria monetarhomoides. Courtesy, Marine Acquarium and Museum, Kavaratti.

The reason why tuna fishing did not develop on other islands in earlier times was because in traditional pole and line method of fishing, the success in fishing depended on access to adequate quantities of the requisite bait fish. These bait fish have very selected habitats located among the reefs surrounding the islands. Minicoy and the Maldives have had a plentiful supply of these fish. Bait fish include *Spratelloides delicatulus, S. gracilis, S. japonicus, Apogen frenatus, A. warrun, A. nigripes, Chromis caeruleus, C. chrysezona, Pterpcaesio tile, P. pisang, Lepid ozygous tapeinosoma*, and *Pempherio owlensis*. Further, before 1960, there was little contact between Minicoy and the other islands and this may have been an additional factor why tuna fishing remained restricted to Minicoy.

Dried *Bêche de mer*/Trepang/Sea cucumber/*Holothuria*, available on both the island groups, was in demand exclusively in China. When the scale of this demand became evident it was transformed into a monopoly in 1836 but by 1881 - 1882 the market had declined and in 1910 the monopoly was abolished.

In 1972-1973 pole and line tuna fishing was introduced from Minicoy to the other islands of Lakshadweep. This could be done, presumably, only after beds of bait fish had been identified by the Department of Fisheries. In present times the most important centres of tuna fishing are Agatti, Bitra and Suheli Par with Minicoy occupying the fourth place on the list. The decline of Minicoy as a centre of tuna fishing constitutes a sad chapter in the history of the islands, and can also be taken as an object lesson on misguided notions of modernisation and development. Circa 1975 the decision was taken to deepen the harbour to attract ships of larger tonnage to Minicoy. This necessitated underwater blasting which led to the mass destruction of *Acropora* corals which formed the habitat of bait fish such as *Bodi* in the southern part of the island. Unhappily, despite this destruction the hoped for diversion in shipping failed to materialise. The Central Marine Fisheries Research Institute at Minicoy has now initiated programmes to try to reverse the trends caused by man-made interference with the eco-systems.

Cowries have also been important to the economy of the islands (*figs. 25-28*). According to Mr. K.C.S. Panicker, Department of Fisheries, Minicoy, money cowrie fall into three categories: *Monetaria moneta moneta, M. m rhomboides and M. m. annulus.* These exist exclusively in coral formations. The closest source for money cowry in the Arabian Sea littoral should, therefore, have been Lakshadweep and the Maldives since earliest times. Since South Asian cowries were in demand in black Africa, it can be assumed that East African sources had yet to developed.

26. Cowrie collection. Chetlat island.

27. Cowrie as life form. Chetlat.

28. Cowrie floor ornamentation. Communal shed for married women for making coir rope. Minicoy.

The house of Arakkal maintained a monopolistic hold over both the coir as well as the cowry trade of Amindivi and Laccadives. With regard to the Maldives, cowries were traded in Africa after 1815 against palm oil and, in Bengal, until the 19th century, against rice. It was perhaps, because of this source of supply of rice that Minicoy and the Maldives were able to avoid the highly inequitable exchange of coir against rice which prevailed in Amindivi and the Laccadives. With the growth of monetisation first in Bengal and then in Africa, the demand for cowries dwindled in the 20th century, its usage being now confined to some parts of Africa where cowries are used in bride price and funerary rites. Since the main wealth of the islands lay in their access to marine life and the by-products thereof, and the luxuriant proliferation of coconut palm trees by which they were blessed, the inhabitants had necessarily to import other requirements. Because of the external demand for *mas* and the seafaring activities of the denizens of Minicoy and the Maldives, they were able to choose their trading partners free of political interference. This was also possible because wants were kept to a minimum and the seduction of luxuries failed to suborn the existing value system.

Aminidivi and Laccadives come within the broad regimen of the winds and currents governing sailing vessel traffic off the west coast of India. Minicoy and Maldives could participate both in the trade of the Bay of Bengal and the Southeast Asian Archipelago as well as that of the Arabian Sea. This factor of wind and weather, irrespective of political control, led to a cultural cleavage between the two sets of islands despite the seeming cementing force of Islam. In the case of Minicoy, in the early decades of the 20th century, boats would leave the island in August with a cargo of coir, coconuts, jaggery and *mas*. The ships would sail to the Maldives and Ceylon where the *mas* would be sold in exchange for items in demand in the Calcutta market. From Calcutta, they would sail to the Arakkan coast where rice would be procured at cheaper rates. After retaining sufficient quantities to cater to the home demand, the remainder of the rice cargo would be sold in the Maldives and Sri Lanka.

Boat Building and Navigational Techniques

Boat Building and Navigational Techniques

The boat building usages and navigational techniques of Lakshadweep can best be understood only if these are woven within the texture of the pan-Indian experience. This becomes all the more necessary because of the interjection of the Wahabi movement into the islands. Not only did the movement empower the less privileged sections but it also brushed aside many past practices smacking of supernatural intervention. It has also been assumed that since the earlier Indian economy was largely self-sufficient, there was no overwhelming reason for the country to develop a strong maritime tradition. However, as early as the Rigvedic period, there are references which point to a knowledge of the sea and the practice of navigation. Some of the rites and superstitions observed by a group of fishermen in the Sundarbans, Bengal, could perhaps relate to earlier practices as well. While at one level, rites pertain to propitiation of the forces of nature, at another, the symbolism is more complex. On the day the keel is laid the caste priest of the owner performs the worship of *bir*, hero/deity. After the boat is ready, the same rite is repeated. Married women, symbolic of fertility, smear the stern and prow with *galai,* vermillion, and mustard oil. Salutation is offered to the boat with *baran dala,* winnowing fan, containing various cereals, a lighted oil lamp, water and *kusha* grass. The presiding deity of the boat is Kastha Devi. There is a notion that the helmsman, *majhi,* should lead an austere life on the boat to earn the blessings of Devi. Devi also provides various indications in dreams but these have to be correctly interpreted. The time for sailing is fixed by the caste priest according to the almanac with the proviso that mornings of Tuesday and Wednesday, are always permissible. Before setting out puja is performed to Sani and Satya Narayan while the caste priest performs that to the Goddess Ganga. At the boat, Ganga, Kali, Kastha Devi, Ban Bibi and Gaji, a Muslim pir may be invoked. Different fishing grounds have different presiding deities who have to be appeased.

There is thus an oral tradition which appears to perpetuate bench marks of an earlier heritage. However, in the absence of material remains any attempt to reconstruct boat building technology as distinct from a seafaring tradition would have to place reliance on the vibrancy of craft transmission both in terms of techniques, processes, tools and implements as well as that of hereditary skills channelised through time across successive generations. The task of identifying the constituents of a tradition which existed at a remote period of time in India is facilitated by the existence of a wide data base of ethnological evidence which lends itself to selective usage. The criteria for selection can be rationally explained as material artifacts and a living craft tradition are both subject to an inner logic which shapes the nature of innovation accepted and the norms which guide selectivity in obsolescence. By virtue of its physical location India has played an important role in serving as an entrepôt between two sectors: Mediterranean Europe, East Africa and the Gulf religion on the one hand and Southeast Asia and China on the other. As a result of this exposure a range of technologies can be identified along the shores of the Indian subcontinent.

South Asia has abutted onto three sets of maritime tradition - the coir sewn tradition of the Arabian Sea inclusive of the East African coast, the *jong* tradition of Southeast Asia impinging on Minicoy and the Maldives, and constructional devices such as the lashed lug. These three constituents could be stated to form the basis of boat building techniques in India no less than the Indian Ocean. The three traditions have not been mutually exclusive. Within the three the outrigger tradition constitutes a subset. The regimen of winds and currents and the struggle for survival in the face of natural adversities has ensured a considerable degree of mutual borrowing and adaptation of technologies among the component maritime groups. Among these adaptations and innovations, mention may be made of constructional usages of Byzantium such as skeleton first (G.F. Bass. F.H. Van Doorninck, *Jr., Yassi Adda, I, A Seventh century Byzantine shipwreck,* Texas, 1982, 311), and the Mediterranean usage of nails (S.Q. Fatimi, *Arab Hulls and Sails*, paper presented at International Seminar, Integral Study of Silk Roads, Road of Dialogue, *Al-Sind and Arab Seafaring, Culture, Commerce and Urbanisation*, Pakistan 27 November-3 December, 1990). Unlike other areas of artisanal activity such as textile manufacture in which similar technologies could emerge in different centres independently of each other, bench marks of boat technology appear to bear a much stronger diffusionist orientation. However, within the geographical spread of a particular technology there appears to be a wide range of region specific variation. This is evident both in the area of the outrigger as well as in that of the coir stitched boat. While the broader parameters of a given technology may remain constant certain constructional usages can reflect local variations in modes of carpentry.

Constructional Methods

While all island crafts are of shallow draft since reefs have to be necessarily negotiated, hydrological and cultural factors have led to a clear differentiation between the Amindivi and Laccadive group of islands on the one hand and Minicoy and the Maldives on the other. This is reflected in traditions of boat building. W. Logan had made this distinction despite the fact that his focus was on compiling the Gazetteer of Malabar. Since the northern and central islands of Lakshadweep had no tradition of fishing on a commercial scale, it was left to Minicoy and the Maldives to develop the specificities of the tuna fishing craft, the *mas odi*. J. Hornell had pointed out this segmentation in tradition in his work. (James Hornell, The Origins and ethnological significance of Indian Boat Designs, *Memoires of the Asiatic Society of Bengal*, VII, 1918-1923, 180-182.)The basic typology of boats, constructed in the Maldives, like those built in Minicoy, have evolved from the peg fastened, lashed lug prototypes of Southeast Asia. There are two kinds of *otis,* the *mas oti* and the *bandu oti*. The *bandu oti* has developed from the *mas oti*. The term *bandu oti* was derived from *bandu*, belly, because the hull was viewed as being shaped in the form of the *bandu*. The *mas oti* is more rounded in form. Despite variations in size, the method of plank fastening is the same in all the Maldivian islands. Prior to 1850 palm leaf matted square sails were in usage. After 1850 Bohra merchants from India began to bring in Indian sail cloth, at which point in time the lateen was adopted.

Circa 1924, R.H. Ellis noted that all boats built in Minicoy were either pegged or nailed. Al Hajjaj (A.D. 661-714) is credited with the introduction of tarred and nailed ships into the Indian Ocean. The Umayyad poet, al-Farazdaq (A.D. 728) refers to ships "fastened by iron clamps". That such boats were not uncommon is attested by the trade in scrap iron obtained from the dismantling of nailed ships which had developed in Basra. However, this reference has no relevance for the Maldivian culture area and the practice of nailing to which R.H. Ellis makes a reference, is to be related to carpentry techniques introduced during the colonial period rather than to any earlier craft usage.

While the emphasis is on the coir sewn tradition some explanation is called for to clarify why the Minicoy tradition differed from that of the other islands. The answer lies, perhaps, in the migratory movements of a group of people, the

Austronesians, who, leaving their home in South China, branched into the Pacific via Indonesia, and to the African island settlement of Madagascar via the subcontinent. This Austronesian diffusion from a common East/Southeast Asian home, may be compared to Aryan, dispersion in middle Asia. Both defy precise definition and the former remains uncharted in relation to movements in the Indian Ocean. There is linguistic rather than archaeological evidence of this group. The usage of the term, *kun lun,* by Wang Gungwu (*Journal of the Malayan Branch of the Royal Asiatic Society, 31.2, 1958, I-135),* could well be taken in the sense of Austronesian. This is the group associated with the cinnamon trade, referred to as *Troglodyte* by the Roman writer, Pliny.

With regard to Austronesian migration to Africa, W.G. Solheim II, has noted that the bearers of cultures who had come to Madagascar were to be found scattered widely over Southeast Asia, Indonesia and New Guinea circa 500 B.C to the year one. (W.G. Solheim, II, Indonesian Culture and the Malagasy Origins, *Taloha*, 1, 1965, 35-36). However, 1000 to 1500 years prior to this date they may have sprung from a common culture base. Those coming to Madagascar could have left their homes in Southeast Asia or Indonesia between B.C. 1000 and A.D. 500, the more probable date lying between B.C. 500 to 1. J. Hornell (*Mariners, Mirror, 1945, 195-217*), was of the view that the route for these migrations may have been via South India and the Maldives. Michel Mollat du Jourdain feels that the most probable course was via the South Indian coast to East Africa and then by way of Grand Comoro to Madagascar (*Archipel*, 21, 1981, 40). B. Domenichini-Ramiaramanana and J.P. Domenichini (*Nouvelles Centres, Art Archéologie,* I, 1983, Madagascar dans l'Ocean Indien pendant le XIII siecle), link Indonesian presence in Madagascar with a conscious effort to widen the network for the distribution of cinnamon. According to these two authors, from the beginning of the Christian era to the 4th century A.D., Indonesia dominated the cinnamon, spice and aromatic trade. Indonesians settled not only in Madagascar but also in Yemen and Arbian Felix. Perhaps Bantu expansion in East Africa, conterminous with Arab expansion in the Indian Ocean, broke the link between Indonesia and Madagascar.

B. Domenichini-Ramairamanana, J.P. Domenichini and J. Innes Miller (*The Spice Trade of the Roman Empire, Oxford, 1969*), have discussed the issue of sea routes and the cinnamon trade in the early centuries of the Christian era in considerable detail. A distinction is, however, made between an eastern group called the Kun Lun and the Indonesians. The validity or otherwise of this remains open. After an examination of the meteorological and hydrographical factors involved, G. Donque (*Taloha,* I, 1965, 42-59), has outlined three possible routes, all feasible to a greater or lesser degree according to the season, linking Indonesia to Madgascar. J. Poirer (*Mouvements de Populations dans l'Océan Indien, Paris,* 1979, 51-59), dates the initial arrival of Indonesians to Madagascar between the third and sixth century A.D. The second phase is dated between the 10th and 12th centuries A.D. For the initial voyages, P. Verin (H.N. Chittick, R.I. Rotberg, eds., *East Africa and the Orient,* New York, London, 1975, 164-191*)*, prefers to identify as the route, the passage along which outriggers are to be found following the southern contours of India and then veering off towards the East African coast. It is to be remembered that both the outrigger as well as the pegged boat tradition had an Indonesian association. During the second phase the route passed through, on the one hand, Sri Lanka, South India, Maldives and Lakshadweep, and, on the other, the Northeast or Northwest coast of Madagascar. Indonesian presence along such an alignment would provide a very plausible backdrop to the presence not only of the outrigger in parts of south India but also of the pegged boat building tradition in Minicoy and the Maldives. Figure 29 captures two faces of individuals on the sea front in Minicoy. An association with Polynesia and Indonesia immediately suggest themselves. It needs to be stated, in this context, that Minicoy and Maldives possess a much stronger seafaring tradition than the remaining portion of Lakshadweep, and even today, memories of past linkages with Indonesia persist in the Maldives.

29. Sea front. Minicoy. Note facial appearance.

The voyage of the "Sarimanok", discussed later, is relevant to this debate, but whether the double outrigger was the boat through which settlement in Madagascar was effected remains an open question. Although familiar in Sri Lanka there is no evidence of the outrigger in Lakshadweep. However, Ali Manikfan, Minicoy, has observed that when children made toy boats in the past, these were equipped with outriggers. Austronesian/Indonesian activities would, however explain the tradition of *mas odi* construction in the Maldives and Minicoy. J. Hornell, (*Mariners Mirror,* 1941, 54-68), also sees a linkage between the *mtepe* of Lamu, East Africa, and the boat building traditions of the Maldives. This is controverted by A.H.J. Prins (*Paideuma,* 28, 1982, 85-100), who feels that the *mtepe* of Lamu and the *pattamar* of the Konkan are both derived from an earlier common prototype. The method used for the joining of planks lends credence to the hypothesis of A.H.J. Prins rather than that of J. Hornell. This would lead to the identification of the Wadiba, the island people, who, according to legend, having been blown off their course, are believed to have settled in Lamu and built the *mtepe* there, with the inmates of Lakshadweep rather than the Maldives as has been postulated by J. Hornell.

30. *Opposite page* - Terrapam. Owner-maker: Hamsath Vellipura, Kavaratti.

Lakshadweep boats can be divided into two categories: those which are rowed and those which depend to much greater extent on wind power. There is also the virtually moribund raft tradition used for subsistance fishing within the lagoon. This latter craft is called *terrapam (figs. 30-32)*, in Lakshadweep, better known through its Tamil Nadu homologue, the catamaran. In Amindivi and Laccadives, the terms, *odi, oti, doni, odam* and *otam* seem to be interchangeable. Smaller *doni* which are rowed have different Malayalam names corresponding to the number of oarsmen on board. The *Valia otam (fig. 33)* sails to the mainland. Earlier, two masted boats with six to seven pairs of rowers, called *kondalam*, existed in Kavaratti. These were mainly used by Thangals to go from one island to another. These were modified to carry cargo at Amini. The generic term developed for these crafts was *ar velikonda* (*figs. 34-35*). Today these do not carry any sail. The *cheriya odi* or fishing boat is shown in figure 36. In all these models the basic method of plank assemblage is that of coir end to end sewing associated with the ships of the Arabian Sea littoral including Western India.

31. Detail of fig. 30. Front portion of terrapam. Note curvature at two end planks to provide a better grip while dragging on sand.

32. detail of fig. 30.Rear end.

33. *Valia odam.* Kalpeni. Note stem post and jutting out lavatory portion at stern. Cooking and living areas are located under the thatched section. Favoured colour for outer hull is black.

34. *Ar velikonda*, Amini. Note that the stem has been wrapped in cloth to reduce wear and tear.

35 . *Ar velikonda.* Purakkara, Amini. Keel slightly narrower at two ends. Note groove on side strake to accomodate cross plank.

36. *Cheriya odi*, fishing boat. Kadmat. The apportionment of space approximates that of the Kondalam. There is no overlap with the Minicoy mas odi geared to the needs of tuna fishing.

39. Coir retting. Each formation denotes separate ownership.

38. Tidal retting of coir. Kadmat. Coconut husk is positioned under sand at low tide being weighted down by coral.

The sequence followed in sewing is the same as that recorded and documented for the East African Coast and Sri Lanka. It would be fair to assume that while the material used for luting and padding, and choice of inner and/or outer hull for application of padding, may be subject to regional variation in the Arabian Sea littoral, the method of sewing remains the same. In earlier times while some timber was imported for building of *Odams* sailing to the mainland, local material was utilised, to a large extent, in the smaller boats. Figure 37 shows the local *Cheerani* tree. *Thespesia populanea,* its naturally curving branches lending themselves to the needs of local boat building.

Apart from requirements of wood, coir rope played a crucial role in the coir sewn boat building practices not only in Lakshadweep but in the entire Arabian sea littorale including East Africa. Figures 38 to 40 show sea shore and pit retting of coconut husk. The positioning of the husks has to be carefully accomplished. Big stones interposed with small ones ensure that neither the husk nor the stones get displaced by tidal movements. The process of decomposition is completed within four to six months. After this the husk can remain in position without any risk of damage. Limited amounts are taken out as and when required. Sea shore retted fibre is regarded as being of first quality while lagoon retted is accorded second place. The superiority of the first process may be attributed to the natural cleansing operations which take place through the surging of the waters in open sea tidal retting.

37. *Thespesia populanea*, the *Cheerani* tree. Kadmat. The natural curvature lends itself to hull planking requirements. The trunk is used for stem and stern. When the tree is reduced to a stump, a fresh growth emerges. Each dwelling has at least one such tree.

40. Pit soaking of husk. Amini. The water in the pit is drawn from the sweet water underground table.

42. Separation of pith from fibre through hand pounding with wooden pellet. Work place of Pura Ahmad Hajji, Amini.

41. Opposite page: Separation of husk from fibre. Husks can serve as domestic fuel. Work place of Pura Ahmad Haji, Amini.

43-45. Twisting of two strands of basic fibre with help of rotating coconut shell .

44.

45.

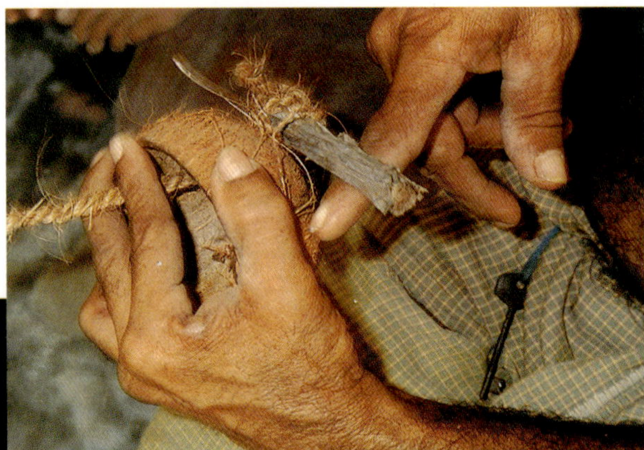

46. Basic unit of twisted fibre in coir rope.

This ensures a more efficient removal of decomposed elements in coarse of retting. Pit soaking provides a relatively weaker product. After completion of the process of retting the husk is dried and fibre separated from the outer covering *(figs. 41, 42)*. This kind of rope which is used as ship's tackle and for sawing planks is called *thuthar*. The derivation is from *thunnuga*, stitch and *thar*, thread. *Thuthar* can be made by either sex. However, in Andrott and Kalpeni, it is a strictly male activity. After separation the fibre is taken to the sea, washed, weighted down lightly with sand and dried on the beach. Usually one day is sufficient. The sand is shaken out and the fibre is brought back. The fibre is drawn out by hand and lightly twisted with the help of a coconut shell *(fig.43-45)*. The initial twisting of *thuthar* is a delicate operation as the cord has to absorb subsequent doubling without snapping. This forms the basic unit in rope making *(fig. 46)*. The thickness of the primary twine varies according to the size of the *Odam* being constructed. The cable is folded into two lengths. One end is held by one person and the second end is knotted around a tree *(Fig. 47, 48)*.

47. Twisting of four units of coir rope around tree. Note rings around trunk of coconut tree. Each ring denotes stage in growth. These also provide footholds in climbing.

48. Detail of fig. 47.

49. Four coils held together in course of twisting. Detail of fig.47.

50. Progression figs. 48,49.

The twisting operation commences. Care is taken to ensure that individual strands are spaced out *(fig. 49)* while twisting is in progress *(fig. 50)*. The four strands are then combined into one (*Figs. 51, 52)*. Figure 53 shows the finished coir rope. For sewing operations, four ply rope is used while for rugging it could range between 8 to 16 ply. The rope for rigging is always made by men. For an *Odam* with a 16 *kol* (40 ft.) length, 16 *thulam* (208 kg.) coir would be used. 1 *Kol* being equal to 2 1/2 ft. and 1 *thulam* to 13 kgs.

51. Twisting of four plies into one unit.

52. Detail of fig. 48.

53. Finished coir cable.
Figs. 41-53 at work place of Pura Ahmad Haji, Amini,

Figure 54 shows the range of tools and implements used in *Odam* construction. All boats are shell built. After planks have been assembled, the first stage is that of working with the *kari*/carbon black *nool*/thread (*fig. 55*). These markings will indicate the mid section of the keel, portions of the hull which will accommodate the curve and points through which sewing will take place (*figs. 56-59*). Only the master carpenter was equipped to use the *kari nool*. At points of curvature the helper would plane down the wood according to directions. Apart from marking the points of sewing the master carpenter would tie the planks together prior to commencement of sewing operations. With the aid of his helper, pegs would be positioned and cloth would be placed for plank joinery. The master carpenter would also mark the position of the mast step for each mast. The drilling of holes and etching of outer recesses was entrusted to an assistant carpenter.

54. Tools and implements used in odam construction. Boat building shed, Kunhi Ahmad, Chetlat.

55. Far right : *Kari Nool* used in marking.

56. String line from stern to stem indicates the centre of the keel. Boat builder, M.K. Kunhi Koya, Malmikakada, Kalpeni.

57. Far right: Additional measurements to check centering in course of construction. M.K. Kunhi Koya, Malmikakada, Kalpeni.

58. Laying down of *Kari Nool.*

59. Lifting of *Karti Nool* after deposit of carbon black. Figs. 58-60 at workplace of U.C. Sayed Ukkaya Chetta, Kadmat.

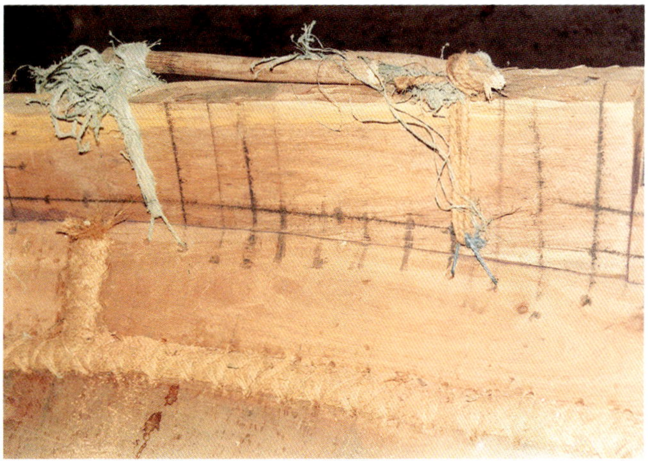

60. *Kari Nool* marking to indicate points for drilling of holes for stitching.

62.

61.

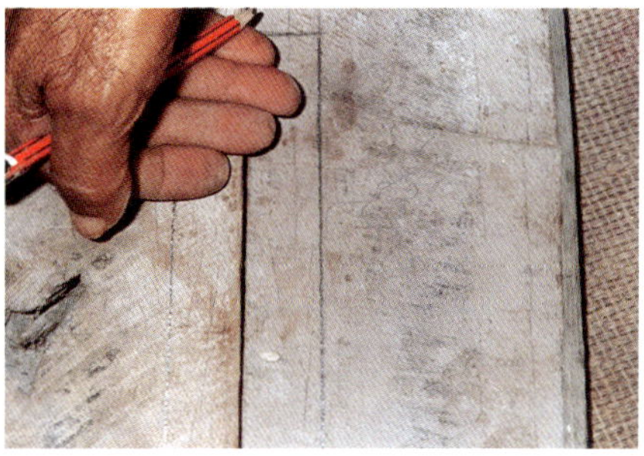

There are fewer and fewer new boats being built today. The majority are refashioned from wood obtained from disused craft. Figure 63 shows one such craft undergoing major repairs. Work practices follow earlier modes with little change in tool or technology. Figures 64, 65, show an abandoned boat in a state of decay. From the point of view of documentation this is of interest as various aspects of the technology involved can be easily perceived.

63. Inner hull in course of boat building and repair work. Figs. 60,63 at work place of Pura Ahmad Haji, Amini.

64. Disused odam on beach. Chetlat.

65. Detail of fig. 64. Note padding between the planks as also along horizontal alignments.

Scarf joining is practiced at the keel *(figs. 66-70)* . For vertical joinery, tongue in groove, *varam patti*, is prefered. Earlier charcoal black was used to ensure a perfect fit. Now this has been replaced by fugitive bleaching blue - Robin Blue. Figure 66 shows one surface which has been smeared with this colour. Figure 67 shows the areas of the upper plank which have picked up the colour. These areas will be planned to ensure a perfect match in surface alignment. Figures 68 to 70, show how the joint is completed. In order to obtain the appropriate curvature at the region of the keel a small section is etched out along the lengths of the two planks which are to be aligned horizontally. The angle between the keel and garboard would measure four finger finger widths (fig. 61). This kind of angular joinery is practiced for laying the first two pairs of planks on either side of the keel (fig. 62). All remaining planks are flush joined.

66. Lower plank in scarf jointing at keel.

67. Method used to identify areas which need planing in upper plank

68. Placement of upper segment in keel joint.

69. Completion of scarf joint. Figs. 66-69 at work place of Kunhi Ahmad, Chetlat.

70. Completion of complex scarf joint. Demonstrated by Kasmi Koya, Pandath, Andrott.

For larger *Odams* pegs may be used for additional strengthening *(figs. 71-73)*. However, these are only positioned along the keel and first two sets of strakes. Five pegs would be positioned on each unit of four meters. It is at this level that the maximum stress is experienced. Pegging does not preclude sewing and serves to reinforce the latter. Pegs may be circular or rectangular. The latter are used for larger odams. Holes for pegs as also for coir sewing are drilled with a hand drill *(fig. 74)*. At the same time recesses are made to accommodate the sewing on the outer hull *(fig. 75,76)*. Figure 77 demonstrates how the hull is kept in position in course of construction. Ribs are constructed as building progresses. Ribbing is, therefore, discontinuous *(cf. fig. 109)*. The uppermost plank is kept in place by temporary bindings during construction *(fig. 78)*. The planks are meticulously planed. Cloth and *pantham*, a resinous sealing substance, are placed between the two surfaces and the planks are then flush joined *(see fig. 75)*. Coir strands are stretched from end to end along the seam of the planks thus brought together *(fig. 79-81)*.

71. Sawing of peg. Mangrove or *Cheerani* preferred. Pegs may be square or rounded.

72. Initial positionig of *ani*, peg.

73. Final placement of peg. Figs. 71-73 at work place of Kunel Shaikh, Chetlat.

74. Drilling of holes by hand drill for stitching. Illustration shows recycling of old planks. Boat building shed of Kunhi Ahmad, Chetlat.

75. Recesses etched to accommodate stitching outboard. Work place of Awwakkal Hamza, Amini.

76. Jointing for stem post. Note placement of cloth between strakes. Work place of Kunhi Ahmad, Chetlat.

77. Positioning of hull in course of construction. Work place of Pura Ahmad Haji, Amini.

78. Detail of fig. 77. Fig. 77, 78 at workplace of Pura Ahmad Haji, Amini.

79. Placement of untwisted coir along seam. Work place of Kunhi Ahmad, Chetlat.

80. Placement of dried coconut husk as in fig. 79.

81. Facing page:Alignment of coir rope over padding. Figs 80,81, workplace of Awwakkal Hamza, Amini.

The sewing operation now starts. For a large *Odam* eight sewers would be required. Four would engage in inboard sewing while four would be positioned outside. Eight sewers would be required for a large *Odam*, four sewing inboard and four being positioned outboard. The needle *(fig.82)* called *aranam*, is made from the spine of a coconut leaf. There is no eye and it is not threaded in the conventional sense. One extremity of the rope is secured by tying. At the other end, where the needle is to be attached, then plies are separated and untwisted. The yarn is divided into two, one part being thicker than the other. The thinner segment is divided into two. The needle is introduced, twisted on and secured by knotting. This knot may be seen in the upper segment of figure 83. The thicker yarn is twisted, pushed through the separated plies of the thinner yarn and secured by a knot. Sewing is done section wise, line by line, as the hull is built up. The master sewer, called *avali panni kar,* takes his position within the hull while his helper sits outside *(fig. 84)*. The first stitch is positioned with a peg inboard *(fig. 85)* to facilitate tightening. The master sewer hammers in the rope *(fig. 86)* while the helper outside tightens the sewing by pulling the pointed peg *(fig. 87)*. The operation is than repeated inside the hull *(fig. 88)* after which the helper pushes the needle to the other side *(figs. 89-90)*.

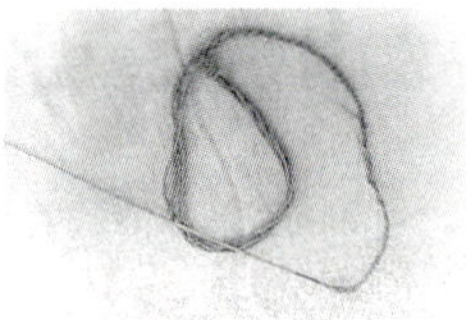

82. Needle *aranam* used in stitching.

83. Joining of rope in graduated thickness to aranam. Demonstrated by Pathada Abossala Haji, Kadmat.

Inboard, the four ply rope is loosened so that it spreads better across the padding. At the initial stage the sewing pattern is V shaped *(figs. 92-93)*. The sewing is completed along the same alignment resulting in a continuous pattern of crosses separated by two vertical lines. The needle penetrates each hole thrice in course of sewing, once in the diagonal plane and twice in the vertical. The only exception is when there is a common hole between horizontal and vertical seams *(fig. 94)*. While sewing is in progress the padding is compacted by hammering *(fig. 95)* as are also the stitches *(fig. 96)*. On completion of the segment the last stitch is kept in place with a peg while the remaining rope is tightened progressively *(figs. 97-100)*. The holes through which sewing has taken place are plugged with coir husk *(fig. 101)*. The needle penetrating to the other side also serves to drag the husk into the hole. The remaining husk is then cut *(fig. 102)*. The sewing hole is again plugged with husk *(fig. 103)* being pried in with a pointed peg *(fig. 104)*, the excess portion being finally cut *(fig. 105)*.

84. First stage in operation of sewing.

85. Positioning of rope from inboard to outboard.

86. Tightening of rope from inboard to outboard.

87. Tightening with pointed peg outboard.

88. Tightening of rope inboard.

89. Pulling needle free outboard.

90. Insertion of needle outboard to inboard.

This operation is repeated until all holes in the given segment have been plugged. On completion of sewing of one segment the process is repeated along the same alignment *(figs. 106-108)* until one sequence along the hull has been covered. As has been stated above, ribs are inserted in a discontinuous manner as the hull is progressively built along the vertical plane. On completion of construction of the Odam *pantham,* earlier used for adhesion of horizontal strakes, is applied on the inner surface.

91. Loosening of ply in rope. Note needle to left above the hands.

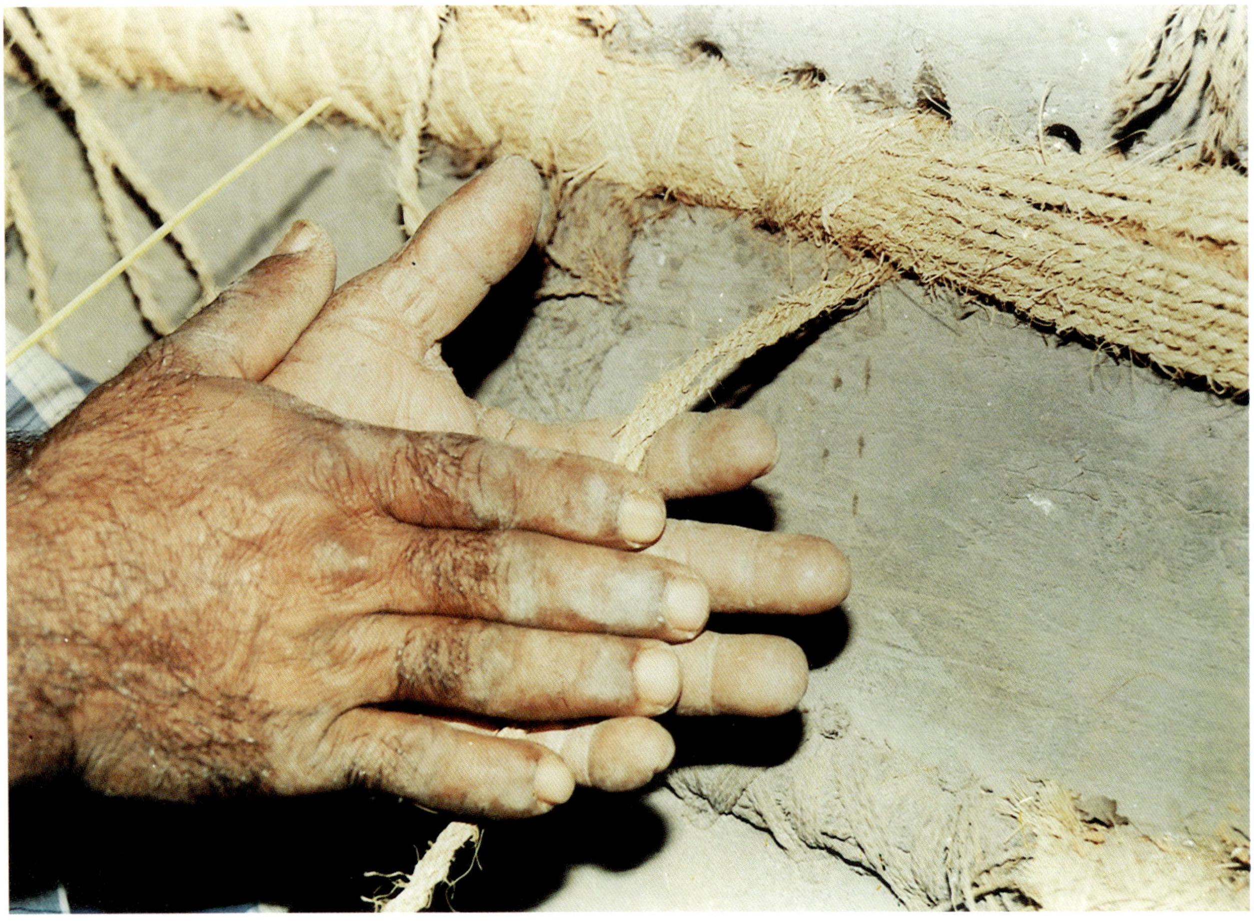

92. Detail of fig. 91.

93. Detail of fig. 91.

94. Common hole through which horizontal and vertical alignement in sewing will take place.

95. Compacting of padded material.

96. Tightening of stitches.

97. Peg to keep stitch in position while the slack of the rope is progressively taken in.

98. Completion of one segment in sewing.

99. Tightening and finishing off of sewing of one segment.

100. Final stage. Tightening at end of one segment in sewing. The process will then recommence from this point.

101. Placement of coir husk in sewing hole.

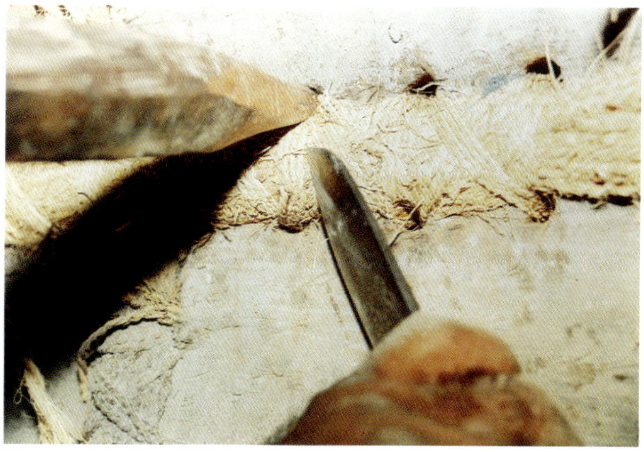
102. After maximum tightening outboard the husk inboard will be cut.

103. Placement of husk in hole.

104. Insertion of coir husk into hole.

105. Cutting of husk.

For the outer surface, lime is added to this solution making the substance more solid. Finally shark oil which is locally available is applied inboard. Figure 109 shows oiling with sardine oil purchased from Cochin. Oiling is done in the sun and after this application the *Odam* is left to dry in the open for four to five days. Once the oil has seeped through the wood it serves as an antifouling agent. The outer hull is sand papered. Finally shark oil to which carbonised and powdered coconut husk had been added is applied on the outer hull giving it a black colour *(cf. fig. 33)*. Although today lines are not as clearly demarcated, in the past carpentry and sewing in boat building were handled by two separate groups of specialists. Plugging for holes and application of *pantham* were also handled by individual groups. The final application of *pantham*, also called *kataram*, and fish oil, *bulakanna*, devolved on the owner who called in the necessary specialists. Mastery in carpentry was accorded on the basis of skill in handling the *karinool*. In sewing, the sewer inboard was the master sewer, the one outboard serving as a helper. A special group among the Malmis concentrated on the cutting and stitching of sails. The metallic needle used for sewing the sail cloth was called *suchi* thus differentiating it from the *aranam* used in plank sawing.

106. First stage in first repeat in sewing.

107. Loose knotted rope indicative of the commencement of the sewing operation.

108. Maintenance of tautness in sewing. Figs. 84-108, workplace of Awwakkal Hamza, Amini.

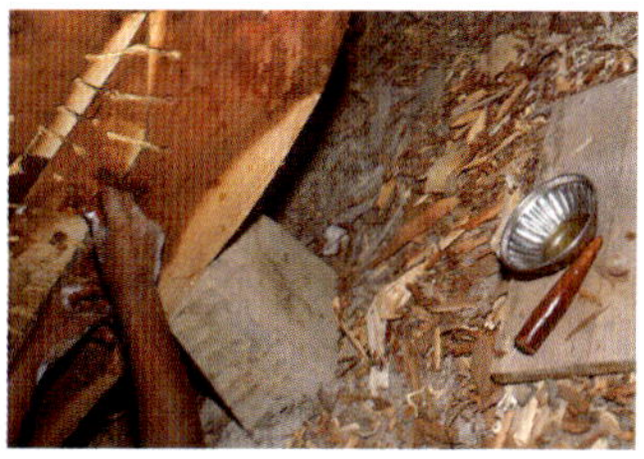

109. *Aruvallikunnu doni*, six oared fishing doni. Oiled with 20 litres of sardine oil. Owner: S.I. Cheriya Koya, Chetlat.

Navigation

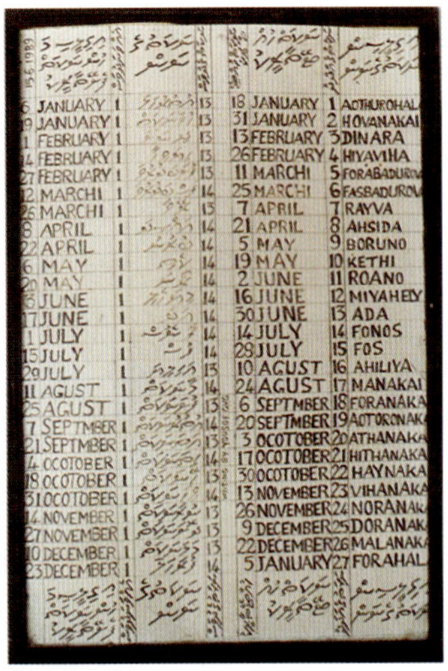

111. Nakai calendar for seasonal activities.. Printed in the Maldives. Based on 27 asterisms. Male Community Hall, *Averhuge*, Falessery, Minicoy.

The knowledge of traditional methods for position finding at sea, still a vibrant component in the oral tradition of Lakshadweep, is an amalgam of the old with a relatively modest input of the new. Navigational methods of finger measurement, *viral*, guide star altitude *(fig.110)*, is demonstrated by K.I. Abdul Kader, Mali/ Malmi of Chetlat. This method of position finding by means of *polegada*, finger measurement, was noted by the Portuguese (Lotika Varadarajan, *Revista de Cultura*, 1991, 198-202). *Kau Kanakku*, learnt by rote and recited in a singsong fashion, lists pole star altitudes at specific positions in terms of a fixed point location in Lakshadweep. *Nakai*, which correlates month, season, star position and favoured activity, is used as a form of almanac in all sailing activity in Minicoy and the Maldives,*(fig. 111)*. Even today there is an astonishing oral repertoire and *ustad malmis*/teacher pilots can slip from Malayalam to Arabic while reciting their codified mnemonic material. These cover aspects such as shapes of star clusters, distances between these, movements of sun and moon and names of *nakshatras* and *rasis*. Such information, chanted with great facility are also called *nelli*. It is conjectured that the system of formulation of astronomical tables, had been initially developed, from Mediterranean usages by the Portuguese . These had been subsequently further refined. This body of material had been grafted by local seafarers onto the indigenous nautical core.

110. The Portuguese referred to angular measurement of star altitude in terms of *polegada*, thumb or finger width measurements to which they were introduced by mariners of the Indian Ocean. K.I. Abdul Kadar, Malmi, Chetlat, demonstrates an earlier practice. The nose is taken as the central point of focus. Star altitude is measured with the laterally placed finger above the fore finger.

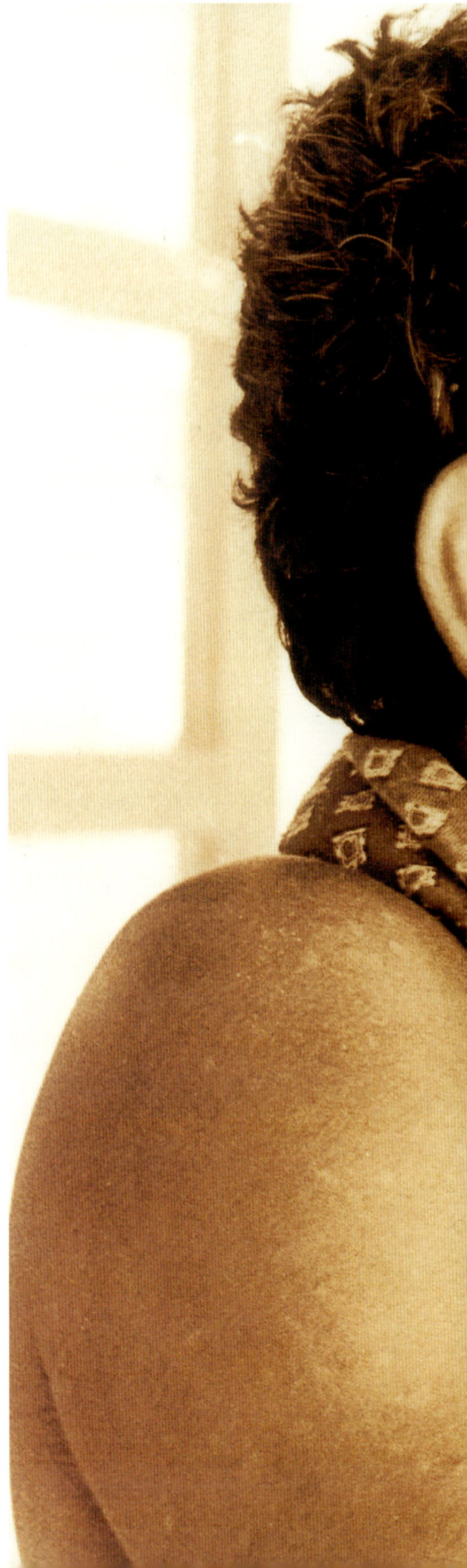

The extant written sailors manuals refer to categories of information encapsulated within the terms *rehmani, daraja* and *nuri*. The scripts include Arabi-Malayalam, Vatteluttu and Malayalam *(fig. 112,113). Rehmani* contain details of latitudinal sailing which are augmented by *daraja*, port location and *nuri*, tables of declination of the sun. *Nuri* tables, introduced circa 1840, represents the European component.. While *kau kanakku* related to latitudinal position finding, the *avavetti kanakku, (fig.112),* like the Arab *tirfat*, introduced correctives for a north-south axis. *Kau Kanakku* which encompasses *anavetti kanakku*, is essentially an oral tradition but because of its centrality to island tradition, it receives mention in local *Rehmani*.

Among the instruments used for measuring altitude mention may be made of the *rappalagai*, the *kaukutty* and the *kolepalaggai*. The *rappalagai (figs. 117,118)* is the *Kamal* to which the Portuguese under Vasco da Gama were introduced and to which they gave the name *tavoletas. (See Obras Completas de Luciano Pereira da Silva*, III Lisboa, 1946, "Kamal, tábuas da Índia tavoletas náuticas", 31-41*)*. The Portuguese suggested an Indian provenance for the *kamal*. Readings taken

112. Local navigational table, *anaveti kanakku*, Malayalam. Invocation on the right reads in translation, "Just as I was blessed by my teacher my blesings go with you". Courtesy Aboobacker, Kavaratti.

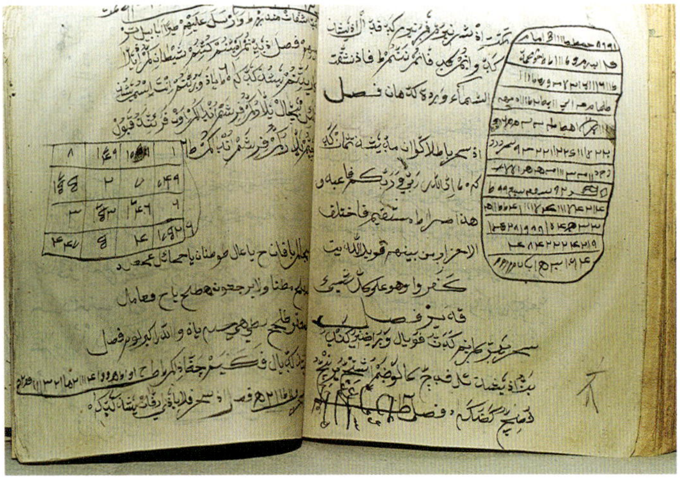

113. Esoteric tables. *Rehmani* of Ahmad Malmi of Malai Ilam, Kavaratti. Courtesy Aboobacker, Kavaratti.

114. K.P. Abdul Jabbar Musliyar, Bandwel. Taking elevation of the sun with the Rubba 'ul Mujayab (quadrant).

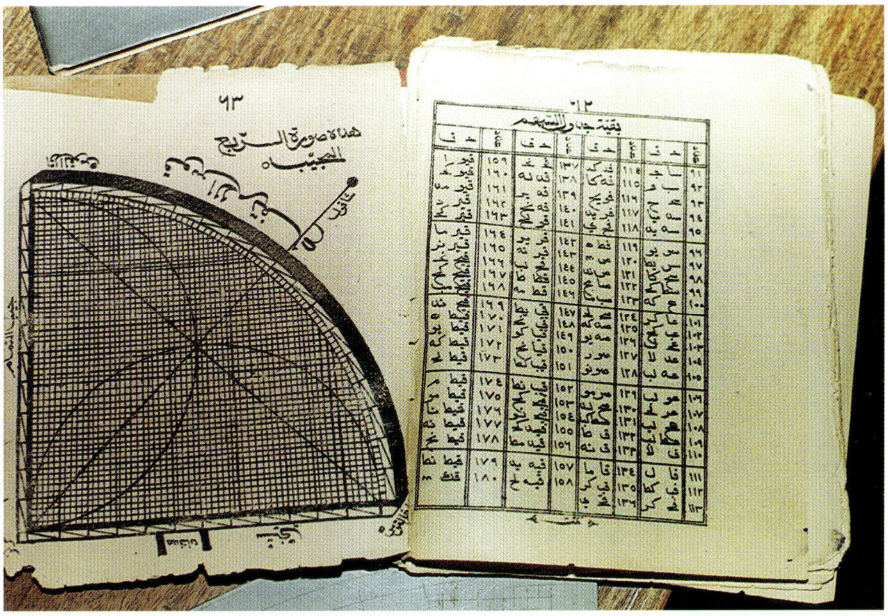

115. Tables to accompany usage of *Rubba 'ul Mujayab*. Courtesy K.P. Abdul Jabbar Musliyar, Bandwel.

116. *Risala*, by Maradim, printed at Tirurangadi, Kerala. The work describes four ways in which the *Rubba 'ul Mujayab* can be used. Courtesy K.P. Abdul Jabbar Musliyar, Bandwel.

by the Portuguese by the *kamal* in the Sea of Oman were incorporated as "Regiment of the Arbalastrille" in the *Livro de Marinhária* by Andre Pires, circa A.D. 1550. The *kau kutty* corresponds to the *goi - daruvan* of the Maldives *(fig. 119).* The instrument called *goi daruvan* in Divehi corresponds to the Arabic, *balishti,* to which the Portuguese ascribed a "Moorish" origin. Adopted from the Orient in the fifteenth century it was progressively improved in Europe in 1606 and 1712. The *kolepalaggai (fig. 120),* in its present form, is very similar to the English backstaff or Davis quadrant said to have been devised circa A.D. 1595. The quadrant, *turiya yantra,* was known in India since the time of Brahmagupta, circa A.D. 598. The Davis quadrant, like the oriental *kamal* and *balestilha,* was geared to the Indian Ocean technique of sailing. This was classified by Europeans as, "running down the latitude". It is, therefore, possible that some form of the prototype of the Davis quadrant may have existed in the Indian Ocean having its origin in the *turiya yantra.* This may have been taken back and perfected in Europe. The quadrant in Lakshadweep is called *rubba 'ul mujaiyab. (fig.114).* Speed of the boat was calculated on the basis of the relation between length of a knotted rope let down, *vegamatra nool,* and time measured through the sand glass *(fig. 121).*

117. *Rappalagai/Kamal.* Tortoise shell. Front view. Courtesy, Aboobacker, Kavaratti.

Usage of the *kamal* would involve computation by means of the *isba* as a unit. One *isba* is equivalent, in Arab usage to a finger or thumb equal to 1.37 degree. The Portuguese used the term *polega/polegada* for this system. Four *isbas* made one *dubban/dhubban*, corresponding to the angle under which one sees the width of the hand when the arm is extended. It is possible that the finger measurements which constituted the *kau kanakku* represented an earlier stage in the development of measurement in terms of *isba (cf. fig. 110)*. This tradition of usage of finger tables could be compared to the methodology pragmatically evolved by the crew for navigating the "Sarimanok". Circa 1985 this vessel, built on the outrigger model portrayed in the sculpture of the 9th century Buddhist monument of Borobudur in Indonesia, sailed out from the Philippines in an attempt to duplicate the course of the earlier wave of Austronesian migration to Madagascar. Unlike India which had developed a highly evolved mathematics based astronomy, Indonesia had no comparable tradition. The "Sarimanok" had minimal instrumentation as the idea was that of attempting to simulate conditions of the earlier period as closely as possible. The navigator, an Australian, evolved his own empirical method, using finger and arm lengths, to judge angular stellar distances for the purpose of direction fixing, hypothesising that such a method could be empirically evolved by early sea people. This would lead back to traditional methods in the Arabian Sea, with the difference that in the latter tradition the erudite had reinforced pragmatic modes.

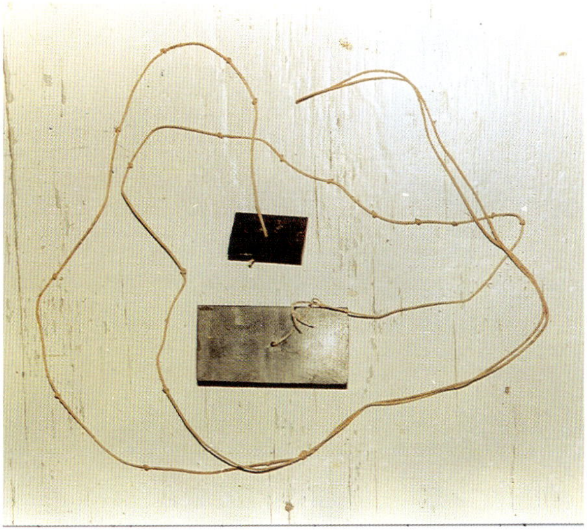

118. As in fig.117. Rear view.

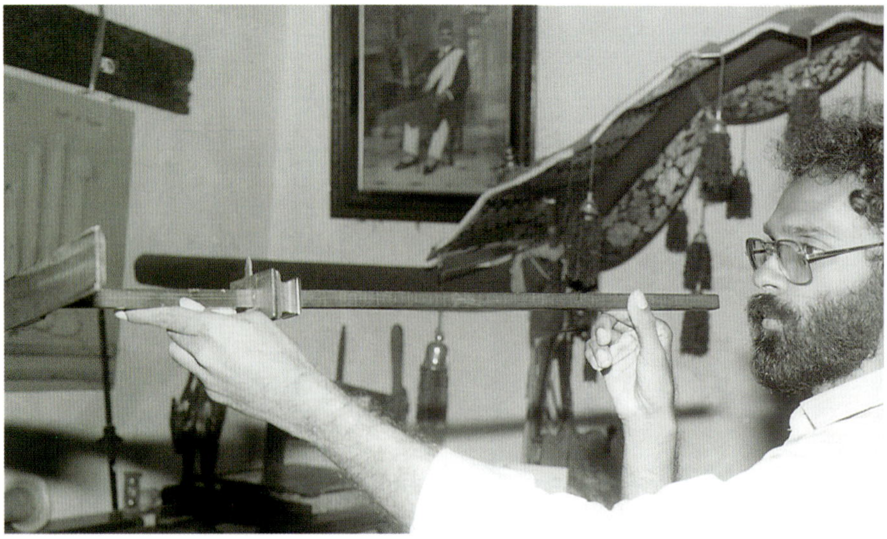

119. *Kaukutty/goi-daruvan/balishti/balestilha*/cross-staff. Courtesy, National Museum, Republic of Maldives.

121. *Kuppi*, hour glass. Used for assessing speed. Courtesy, Aboobacker, Kavaratti.

120. *Kolepalaggai*. (Davis quadrant in European terminology.) Kiltan.

Conclusion

Despite the importance of the coir and cowry resources of Lakshadweep for the Indian subcontinent, very little is known about the early history of human settlement on the islands. While economic links may have brought the islands closer to the subcontinent, environmental and hydrographic factors opened other channels of interaction and exchange to the islanders.

Austronesian impact has yet to be analysed and it is to be noted that despite the caste-like structure of social organization of the islands which point towards contact with South Asia during its Hindu - Buddhist phases, acceptance of Islam by the islanders changed the context. It provided easier access to the trade network of the Arabs bringing with it possibilities of freer interchange in areas of nautical technology and associated usages. Arab nautical technology, subsequently codified, was a composite creation comprising many elements of the preexisting South Asian repertoire. There had been similar hybridization in Chinese boat building and nautical tradition consequent to its opening to Indian and Indonesian maritime expertise. The sailing vessels of Lakshadweep and the Maldives could provide links in this chain of transmission which spanned the region stretching between East Africa and Indonesia.

In consonance with the South Asian attitude to adoption of new technologies, islanders were also governed by the concept of *sufficiency*. This meant that an existing technology would absorb additional grafting until a point of saturation had been reached when the system would no longer suffice to serve the felt confluence of needs. It was at this juncture that qualitative change could be introduced.

The adoption of instruments such as the compass and sextant *(kaman)*, radically displaced earlier approaches to navigation in Minicoy. There is a wealth of astronomical, astrological and navigational data in different media to be found in all the islands. This includes innumerable sailing manuals, wood carvings and standing stone tablets. Apart from this there is also the richness in terms of human assets which encompasses knowledge of wind, wave, current, ocean floor as also non-esoteric Thangal learning in the Tarwad islands. All this tends to controvert the stereotype that the best in existing Indian naval tradition can be traced solely to the colonial experience.

The chief *arkati* of Minicoy (*fig.123*) had to have complete mastery over knowledge of hydrographic factors as well as of the topography of the intra-lagoon sea-bed. This allowed little place for the supernatural. At the same time, while the fisherfolk and the average seafarer may have been governed by empirical traditions of navigation, the great *malmis,* and *malmi* ustads/pilot teachers were literate in Arabic and had access to theoretical and scientific as distinguished from *technological* traditions. When it came to nautical material in English, as evidenced by the Portuguese mariners of yore in their encounters with local pilots in Asian waters, the compendium of knowledge at the disposal of the Lakshadweep mariners enabled them to surmount the language barrier and grasp the essentials. This is in clear in the area of *Nuri* derived knowledge.*(Fig. 122)*

123. Ali Thakru, *arkati*, chief pilot and harbour master. Edulkage, Minicoy.

With regard to the shipbuilding tradition of the Indian Ocean it is to be noted that certain features in construction remain constant despite the play of other variables. These can be taken as the benchmarks of a particular tradition. If, in a given region, the basic method of plank joinery remains the same, with departures in rigging and hull, it is the first which would impart specific ethnicity and constitute a unifying feature in groupings postulated. This has to be borne in

122. Instructional material inscribed on wood. Appear to be based on Admiralty charts. Teaching aids for *malmis*/pilots. Probably devised by Kunnamkalam Muthu Koya Tangal, Andrott. Courtesy, Mapilate Pukoya, Andrott.

mind with regard to sewn boat typology within such a connotation. The method of coir stitching of planks cannot be accommodated within the two categories of luting and caulking posited by B. Greenhill *(Archaeology of the Boat,* London, 1976). The padding procedure, which completes the sewing process, approximates caulking more closely than luting. Perhaps *stitching* needs to be constituted as a third category in typology of plank joinery.

Additionally, the body of knowledge encompassed in *Kou Kanakku* suggests interesting possibilities. Arabs followed similar graduations in their own series of finger tables. The point of origin of these tables is unknown although it is suggested that this may be ascribed to some point in the regions abutting on 10° North latitude. This would include Lakshadweep within its sweep. However, geographical location alone cannot provide sufficient basis for the islands to merit inclusion for putting forward such a claim. The attainments of Lakshadweep in the areas of seafaring and navigation do tend to buttress this position. In this context, it is important to remember that Greek attainments in instrumentation could be accessed without interruption by India since the time of the invasion of Alexander of Macedon. Europe, however, had to wait until the fifteenth century to do so. The early mention of the quadrant in Indian texts and its usage in the islands (cf. Lotika Varadarajan in A.T. de Matos, L.F.F.R. Thomaz, eds., *A Carreira da Índia e as Rotas dos Estreitos,* Actas do VIII Seminário Internacional de História Indo-Portuguesa Angra do Heroísmo, 1998, 690-694), is one instance of such contact. The story of Lakshadweep has to be probed much further but the curtains appear to be slowly lifting.